The Colouring Guide

BIRDS
AND
BUTTERFLIES

Piccolo
A Piper Book

Woodlands

The woods are full of bird noises. As well as the many different types of song, you may hear the tapping of a Woodpecker, the flapping of Pigeons in the treetops, or the sound of small birds scratching in the undergrowth for food. Many birds build their nests high up in the trees, or in tree hollows, where they are safe from their enemies and sheltered from the weather. Others nest on the ground, hidden among the leaves. As well as offering protection, the woods also have a huge variety of buds, nuts, seeds, leaves and insects for birds to eat.

Among the most beautiful of the insects that feed on the trees are the butterflies. They are also attracted by the colour and scent of the many wild flowers that grow in the clearings of the forest.

(The various parts of a bird are shown on the inside back cover)

Woodpigeon (41 cm). The plump Woodpigeon is a bluish-grey colour, with a white patch on both sides of the neck and a white band across the wings. It builds its thin nest of twigs in treetops and high hedges. In the daytime, large flocks of Pigeons leave the cover of the woods to feed in nearby fields and meadows, where they can do much damage to farmers' crops. They can be seen all year round.

Bullfinch (15 cm). Bullfinches are often seen in pairs, as they are said to mate for life. The male has a bright red front and grey back, while the less colourful female has a salmon-pink breast and brown back. Both are easily recognized by the white rump, which may be glimpsed as they flit through the undergrowth or along a hedgerow. Bullfinches are found all year round in woods and orchards, where they raid trees for seeds, fruit and flowers.

Green Woodpecker (32 cm). The loud, laughing call of the Green Woodpecker can often be heard before the bird is seen. Both male and female are green, with a yellow rump and red crown. The female has a black cheek stripe, while the male has a red stripe with black borders. It picks insects from trees, and carves a round hole in the bark for a nest. It can also be seen on the ground, hunting for ants with its strong pointed bill and long tongue.

3

Redstart (14 cm). This brightly-coloured small bird is one of the earliest visitors to arrive in Europe from tropical Africa in the spring. The male (above) is a striking combination of colours, with his grey crown and back, white forehead, black throat and red breast. The female is browner and paler, but both have a red tail which they flick up and down.

Cuckoo (33 cm). The song of the Cuckoo is one of the first sounds of spring. It arrives in Europe in April, returning to Africa in August. Both male and female are mainly grey, with bars on the white underside and a long, spotted tail. The female does not make a nest, but lays her eggs in the nests of small birds such as the Pipit or the Reed Warbler. The baby Cuckoo grows quickly, pushing out the other eggs and clamouring noisily for food from its tiny foster-parent.

4

Speckled Wood (35–45 mm).
This butterfly can be seen between
April and November, in shady
woodland clearings or among the
hedgerows. Its colour-pattern of
creamy-white markings with
round 'eye-spots' on brown wings
makes it difficult to see against the
dappled light of the woods.

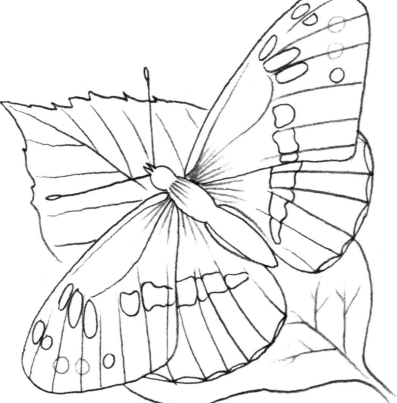

White Admiral (52–60 mm).
The White Admiral is found in
mixed woodland, and is often seen
around Bramble flowers. Despite
its name, it is mainly dark brown,
with white markings and two
rows of black spots on the
underside. It is seen during the
months of June and July.

(The various parts of a butterfly are shown on the inside back cover)

Silver-washed Fritillary
(56–74 mm). This is quite a common woodland butterfly, seen between June and September. It is brownish-yellow in colour, with black markings on the forewing and hind wing, and striped silvery markings on the underside.

Fields and Meadows

Some of the birds which live in the woods come to feed in open countryside during the daytime. They return to the nearby trees at night, and it is there that they build their nests. Others spend only the warm summer months in fields and meadows, and fly south to hotter countries for the winter. However, many birds make their homes in the fields, hiding their nests among the grasses or in the hedgerows. Fields sown with seeds and grain attract large flocks of birds, and they can become a menace to the farmer. But many birds feed on weeds or meadow grasses, and some prefer a diet of worms, slugs and snails.

Many butterflies feed and lay their eggs in meadows. The tiny caterpillars which hatch from the eggs grow quickly, and they eventually spin a cocoon called a chrysalis. Inside the chrysalis, the caterpillar changes form, and at last breaks out as a butterfly.

Skylark (18 cm). The Skylark can be seen in all types of open countryside. The female makes her cup-shaped nest on the ground, where her speckled brown and white colouring makes her hard to spot. The Skylark has a clear, warbling song which it can keep up for many minutes as it swoops over the countryside.

Goldfinch (13 cm). The bright colours, dancing flight and musical song of the Goldfinch make it one of the most charming birds of the countryside. It is easily recognized by its red, white and black head, black wings with wide 'gold' band, white rump and black tail. Goldfinches are usually seen looking for seeds among trees and bushes, or in fields of thistles.

Yellow Wagtail (16 cm). The Yellow Wagtail is a slender bird with a long, constantly bobbing tail. It has bright, golden-yellow underparts, and an olive-green back. It arrives from Africa in April, and spends the summer months hunting for insects among marshes, fields and water-meadows.

Pheasant (Male: about 82 cm; Female: about 58 cm). Pheasants can be seen on the ground by day, scratching in fields for seeds, grain and insects. They roost in trees, and may also be seen perching there in the daytime. The colouring of the male varies, but the dark green head with red eye patch is typical. Some, like the one in the picture, have a white collar. The female is smaller, with a speckled brown body.

10

Marbled White (45–56 mm). The Marbled White butterfly is a familiar sight in open fields and meadows, where the caterpillar feeds on grasses. The butterfly, which is a creamy-white and brown colour, can be seen in June and July.

Dark Green Fritillary (48–65 mm). The only green colouring on the Dark Green Fritillary is a dull patch at the base of the wing. Otherwise, it is reddish-brown, with black markings. It is seen in July and August on flowery slopes, heaths and hillsides, where it is particularly attracted to purple flowers.

Clouded Yellow (47–61 mm). The yellow wings of this butterfly have black borders, with a black spot on each of the forewings. The female has similar markings, but also has yellow spots in the black borders. The Clouded Yellow is seen from early spring to October, in fields of Clover or Lucerne.

Small Copper (25–30 mm). A small and lively butterfly which can be seen between May and October in dry fields and heathlands. Its forewings are a coppery-orange, with brown borders and markings, while the hind wings are brown with orange borders.

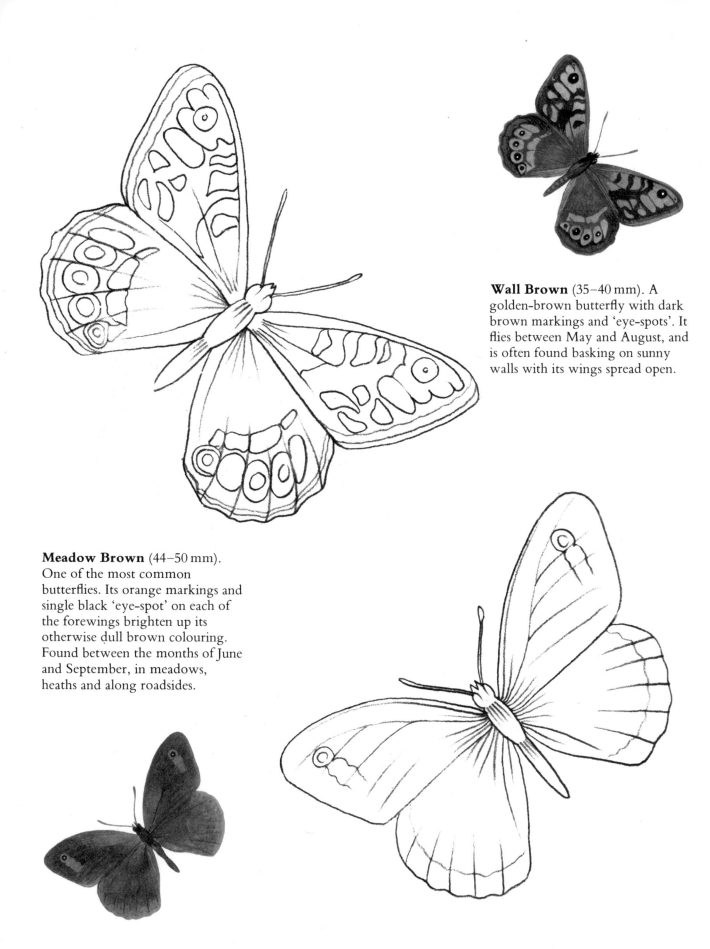

Wall Brown (35–40 mm). A golden-brown butterfly with dark brown markings and 'eye-spots'. It flies between May and August, and is often found basking on sunny walls with its wings spread open.

Meadow Brown (44–50 mm). One of the most common butterflies. Its orange markings and single black 'eye-spot' on each of the forewings brighten up its otherwise dull brown colouring. Found between the months of June and September, in meadows, heaths and along roadsides.

Swallowtail (60–98 mm). A large black and yellow butterfly, with blue borders and a red spot on each of the tailed hind wings. It has a graceful, swift flight, and can be seen between April and August in marshes and flowery meadows. In Britain, the Swallowtail is only found in the fenlands of East Anglia.

Parks and Gardens

Towns provide food and shelter for many different types of bird. Even in city centres, large flocks of Pigeons and Sparrows nest on buildings, and feed on the scraps of food that are dropped by people.

A greater variety of birds can be seen in parks and gardens. Tits hop through trees and shrubs in search of insects. Thrushes feed on worms and snails, and Finches crack open the seeds of weeds with their strong bills. Birds will use all sorts of perches for a song post, such as chimneys, television aerials, fences or washing lines. And many will visit bird tables, especially in the winter when food is scarce.

Some of our most colourful butterflies can also be seen in parks and gardens, where brightly coloured and strongly scented flowers provide plenty of nectar. And caterpillars can often be found feeding on the leaves of stinging nettles around wasteland and building sites.

Greenfinch (15 cm). A stout, short-tailed finch with a powerful bill for cracking seeds. The green-brown body is brightened by flashes of yellow on the wing and tail, although the female is paler and browner than the male. Greenfinches are often seen in garden trees and bushes, and may also be found feeding among the weeds on wasteland.

Great Tit (14 cm). The Great Tit is a familiar sight in towns, where it can often be seen hanging from garden bird tables or boldly pecking at the tops of milk bottles to get at the cream. This largest member of the tit family has a black cap and throat, and a black stripe down the middle of its bright yellow breast.

House Sparrow (15 cm). The most commonly-seen town bird, found in city centres as well as in parks and gardens. Sparrows have a streaky brown back and a stout bill for eating seeds. The male has a grey crown and a black throat, while the female is lighter brown with a pale eye stripe. They are seen at all times of the year.

Blue Tit (11 cm). Smaller than the Great Tit, the Blue Tit has a blue cap, wings and tail and a bright yellow breast. It is another popular visitor to the town garden, especially where bird tables are hung with nuts and fat.

Robin (14 cm). The plump, long-legged Robin is one of the best-loved garden birds. It is easily spotted, with its bright red face and breast, and brown back. It is usually seen alone or in pairs, since it guards its territory jealously from other Robins.

Song Thrush (23 cm). The Song Thrush is a brown bird with a pale, spotted breast. It is famous for its song, which it repeats many times over from a high perch, such as the treetops. It feeds on fruit, berries and insects, and can often be seen smashing snail shells against a stone or pathway.

Red Admiral (58–69 mm). The Red Admiral is a handsome dark brown butterfly with red, black and white markings on the forewing. The hind wings have a red border, with a blue patch at the base of each wing. It is a common sight in gardens between May and October, being attracted to flowers and fruits.

Small Tortoiseshell (44–52 mm). The Small Tortoiseshell is one of our best-known butterflies, with its stripey orange, yellow and black wings with blue border markings. It can be seen as early as March, through till October. It is often found on plants which are rich in nectar, such as Buddleia (shown below).

Peacock (56–68 mm). The beautiful markings of the Peacock may be used to frighten away enemies, as when this butterfly opens its wings it shows four large 'eyes'. It flies on warm days in early spring, but is seen mainly during May and June.

Ponds and Riversides

Not all the birds which feed at ponds and rivers live on the water itself. Some swoop above its surface, catching insects on the wing. Others hide among the reeds at the water's edge, and are difficult to see. The Kingfisher builds its nest in the river bank and hunts from nearby trees, diving into the water to spear a fish.

However, a great number of freshwater birds have special features to help them live on water. Ducks, swans and geese have webbed feet, which they use as paddles to swim strongly. Their bills are broad and powerful, to help them pull up plants and grasses. Dabbling ducks, like Mallards and Teal, feed on water plants and tiny animals on or near the surface, while others dive underwater for food. Many of these water birds can be seen on park lakes in towns, as well as in the wild, and most have become quite tame.

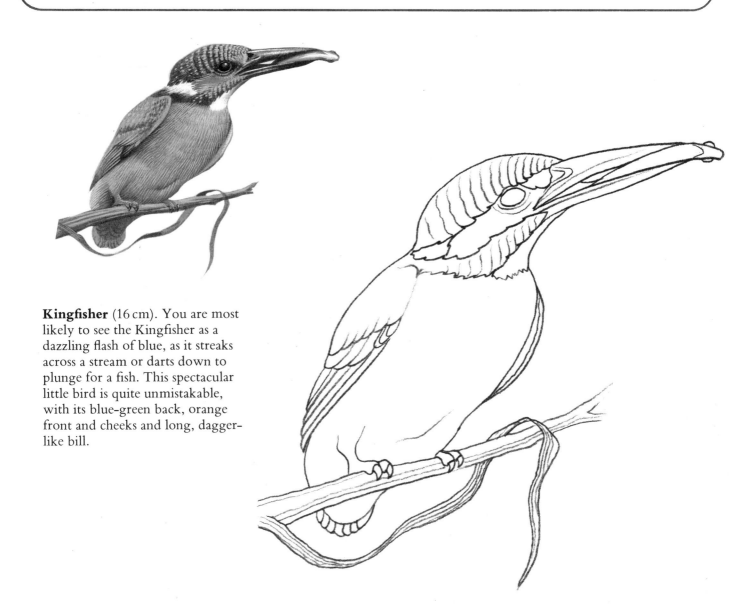

Kingfisher (16 cm). You are most likely to see the Kingfisher as a dazzling flash of blue, as it streaks across a stream or darts down to plunge for a fish. This spectacular little bird is quite unmistakable, with its blue-green back, orange front and cheeks and long, dagger-like bill.

Mallard (58 cm). The Mallard is our most common duck, being found in town parks, country lakes and river estuaries. The male has a glossy green head, maroon breast and narrow white collar, while the female is brown. Both have a blue wing patch. They are dabbling ducks, feeding on plants on or just below the surface of the water.

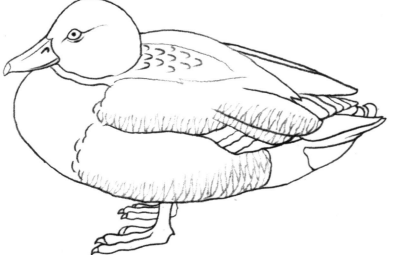

Teal (36 mm). A very small dabbling duck. The male has a broad green eye stripe edged with yellow, on a chestnut head. The female is brown. Both have a green and black wing patch. They are agile little birds, and will spring directly into the air from the water if alarmed.

Shoveler (51 cm). The Shoveler is found in ponds and marshes, where its unusual spade-shaped bill helps it to strain plants and tiny animals from shallow waters. The male has a green head, white breast, blue and green wing markings and chestnut flanks. The female is a speckled brown, with blue and green wing patches.